The Global Superorganism

Jens Zett
c/o AutorenServices.de
König-Konrad-Str. 22
36039 Fulda
jens_zett@protonmail.com

All rights reserved

ISBN-13: 978-1539530176
ISBN-10: 1539530175

Inhalt

I. Preface 6

II. The development of life on Earth 9

III. Biological Superorganisms 16

IV. Superorganisms as evolutionary driving force 23

V. A global Superorganism 32

VI. A view on the World 40

I. Preface

Life on earth is a mystery. The wonderful diversity of lifeforms, their baffling emergence and the incredible sophistication of Nature itself, have amazed and fascinated mankind since its beginnings. Resulting from this fascination a series of theories on the formation and deeper meaning of life have emerged, but due to the very limited knowledge of mankind during the most part of the last few millennia, only very shallow theories could be constructed, which often relied on a greater power to explain unknown things. Consequently, a greater being was often used to explain the creation of the universe and life itself. Remnants of these theories are still found in most religions existing nowadays, which have conserved these kinds of theories from the ancient times. However, human knowledge has evolved due to massive scientific breakthroughs over the last centuries, which has led to new and improved theories in every scientific area which can explain the observable world better than their predecessors.

Oftentimes, the new theories did not render the old ones obsolete but rather enhanced them: Newton's mechanical theorems were precise for small velocities, but the more the speed of a body approached the speed of light, the more Newton's predictions would diverge from the observations. Einstein fixed this discrepancy by enhancing the Newtonian mechanics with his theory of relativity, which could describe and predict the real world better.

Likewise, theories on the origin of living beings have been refined in the last millennia. Ancient theories acted on the assumption that the world was static, since due to the lack of an objective observer, the subjective feeling was that the world does not change. Over time, it became clear, that this was not the case at all and that, although changes during a lifetime were often too little to be perceived, it could be shown through scientific advances, e.g. in geology or archaeology, that Earth is extremely old and life forms have changed dramatically during these gigantic time periods. These new insights were then used to construct new theories to explain the world better, for example by Lamarck or Darwin.

To the present day, a slightly modified version of Darwin's theory is still widely accepted as correct. This theory states that the evolution of life forms on Earth are caused by random mutations in conjunction with a natural selection. Random mutations in the genome would therefore occur by chance because of natural circumstances, e.g. radiation. A mutation can either enhance the probability of survival and creating offspring, or decrease it; it is therefore either positive or negative for the life form. Natural selection then makes sure that the positively mutated life forms have a greater chance of survival and procreation than the negatively mutated ones. This in turn would cause the positive mutated life forms to outcompete the negative ones and over time the good mutation will spread throughout the population.

A relatively simple theory, which tries to explain far reaching phenomena with very few assumptions. Although parts of Darwin's theory are still definitely true today, e.g. the conclusion that all life evolved from a shared ancestor, there are "new" scientific observations that poke big holes in

Darwin's theory of random mutation and natural selection [1] [2]. Especially the exponential advancement of devices which are used to analyze the world from the nanoscale upwards, as well as a continually deeper understanding of genetic processes, expose the limits of Darwin's theory. The point of no return, at which Darwin's theory cannot implement new insights, has been reached a long time ago, even though the scientific mainstream has failed to recognize, or chosen to ignore it to this day [1] [2]. However, it has become apparent that a new theory of the Evolution of Life is needed to explain the observed reality.

II. The development of life on Earth

The biological organisms living on Earth today are unimaginably complex. The high complexity and virtually perfect adaption to their respective habitats, may lead many people to the conclusion, that these life forms had to be created "such as they are" by an all knowing and almighty "God", as is argued by many religious institutions. However, it has become clearer and clearer since humans began exploring the age of the Earth scientifically, that it was much, much older, than previously assumed by ancient theories. According to present knowledge, the Earth is around 4,6 Billion years old and has been harboring life for about 3,5 to 4,1 Billion years [3] [4]. This is of course much longer than the few thousand years which some religious philosophies assume the Earth is old. But even centuries ago, geologists could deduct from the existence of different rock layers that the Earth had to be very old and was subjected to massive environmental changes during its existence. The fossils that were dug up, told a story of perpetual changes in living organisms, which made it apparent that a massive developmental process has taken place, with numerous diverse life forms and intermediate stages between them, from which all present animals, as well as Man itself, originate. These insights of course sparked resistance by advocates of outdated beliefs, but the perseverance of some scientists paid off and the Evolution of Life is widely recognized as true today. However, even though the existence of an evolutionary process can clearly be observed, the mechanisms are hardly understood as of today. The present standard theory of Evolution, which is still seen as the "gold standard" is Darwin's theory on the origin of species. According to Darwin, the evolution of life forms is relatively simple: Random hereditary mutations occur, which can be either good or bad for the organism in its present habitat. The organisms with beneficial mutations have therefore a higher chance of survival and procreation, which causes the good mutations to prevail and spread through the population. Accordingly, new species develop if two populations of the same species are separated in different locations and divert from each other genetically because of different random mutations. If the populations have become too genetically different, they can no longer procreate with each other and therefore belong to two different species.

This theory seemingly provides everything that is necessary to explain Evolution: The emergence of mutations in organisms (random mutations in the genetic code), the improvement of species (because good mutations are naturally selected), as well as the development of new species (spatially separated groups develop different mutations, until they are no longer the same species). But there are a number of findings and insights that contradict these hypothesized mechanisms of evolution: An evolutionary process, which underlies Darwin's laws, would be steady [1], since statistically, random mutations always have the same chance of occurring. Also, these random mutations would be spread evenly across the genome [1]. This is comparable to rolling the dice, where each number has an equal chance of being on top, which results in an even distribution over a big number of casts. The observed mutation rates however, are neither in space nor in time evenly spread [1] [5] [6]: The past has shown that massive evolutionary leaps happened in short periods of time, during which the genome was changed dramatically [7], while the time between is characterized by a genetic stasis, during which hardly anything changes [1] [8] [9] [10].

However, even evolutionary leaps did not change the whole genome, instead some areas underwent massive changes, while others did not change at all [1] [5]. These observations about the spatial and temporal inhomogeneity in evolutionary advances should make one wary about the integrity of Darwin's postulated evolutionary mechanisms of random mutation and natural selection.

The underlying genetic processes that led to the discrepancies in Darwin's theory have been known for quite a long time; however the consequences of these findings have either been ignored or have not been acknowledged yet. The genetic changes in specific areas at certain times are carried out by Transposable Elements. Also called Transposons, these are distinct sections on the DNA string, which are some kind of genetic "toolset", that can, simplified, cut, copy, insert, inactivate or delete genes [1] [5]. On its own, the existence of such tools should have led to a challenging of Darwin's theory, since mutations do not exclusively happen randomly, but can be carried out by the tools the genome itself contains. The knowledge of these Transposons is hardly new, Barbara McClintock discovered them in the 1950s in Maize [11] [12]. Since the discovery that the genome can change itself contradicted the widely regarded as right theory of Darwin, McClintock was plainly ignored and the results of her experiments hushed up. It wasn't until 1983, that the correctness of her experiments and conclusions was acknowledged and she received the Nobel Prize of Biology, after many scientists published similar findings. Still, Darwin's theory was not adjusted to fit these findings since then [2]. Even after the release of the human genetic code by the Human Genome Project in 2003, the theory was not challenged. This is surprising, since the genome did not contain, what was widely expected: Namely, that the whole DNA consists of protein-coding genes. In reality only about 2% (!) of the DNA consists of protein blueprints, while Transposable elements make up about 40% [1] [13]. These were short handedly declared as "Junk DNA" that randomly accumulated and did only serve its own purpose (egoistic genes). But over time, the real function of these often repeated elements was discovered and the "junk" hypothesis is seldom represented anymore. It became clear, that the DNA is not just a collection of protein blueprints, but instead some kind of database with executable "programs". Eshel Ben-Jacob coined the term of "cybernetic agents" and declared the genome as some kind of "computer" [2].

The existence and ability of transposons to change the DNA could explain why evolutionary changes occurred only at specific times in specific areas of the genome: During times of genetic stasis, the transposons were inactive, so only random mutations occurred, which hardly have evolutionary impact. Depending on how the genetic tools where activated, different parts of the genome could be changed.

According to Darwin's evolutionary mechanisms, all life forms would be exposed to a similar number of random mutations and therefore to a similar evolutionary rate. Observations however have shown that this is absolutely not the case: There are species which have not changed much, if at all, for hundreds of millions of years, for example Crocodiles or the Nautilus, while others like mammals have evolved tremendously in the same time span.

It may seem baffling, that in spite of the many observations which speak against Darwin's postulated evolutionary mechanisms; they are still widely accepted as true. However, this also has historical and societal reasons: Darwin devised his theory of the evolution of life explicitly without the influence of an unknown "force" as to keep the catholic church from regaining influence on science [1]. Making chance the evolutionary driver was a very clever move in this regard. Still Darwin said, random mutations should only be regarded as random, until their true origin was discovered [14]. But even today, there are numerous efforts of religious extremists to try and get archaic religious theories taught in schools, especially in the scientifically very advanced USA. In spite of the university driven science being world class, a very strong movement of creationism or "intelligent" design exists, that perpetuates the notion that life on Earth was created exactly as postulated in the genesis saga of the bible and this should be taught as fact in schools. It may seem absurd that these "creationists" use seemingly scientific arguments to discredit the Theory of Evolution, while an even rough scientific examination of the genesis saga debunks it as fiction. But the irrational momentum of this movement made many scientists wary of criticizing aspects of Darwin's Theory of Evolution, as not to provide a point of attack for creationists to gain influence on science itself [15]. It is feared that the creationism movement will become stronger, if the scientists do not stand united against them. Still, it will get harder and harder over time to defend the Theory of Evolution since many inconsistencies with current observations and theoretical predictions have come to light, which creationists could use to discredit the scientific viewpoint of evolution itself. It could be argued, that the Theory of Evolution as a whole must be wrong, since the postulated mechanisms are obviously false according to new scientific findings. It is absurd that people, who believe in an utterly unscientific genesis theory, could undermine a scientific theory with scientific arguments. Without a revision of the evolutionary theory this might very well happen.

For this reason, it is of utmost importance, even under the threat of the invasion of science by religious viewpoints, to be open for new observations and always challenge current theories critically. Random mutations as the absolute evolutionary force have become obsolete by the observation of genome changing tools. Since transposable elements can deactivate, cut, copy, insert or delete genes, evolution is not dependent on random mutations. Critical is the coordinated activity of these tools, since they are under normal circumstances very much inhibited by processes in the cell, which results in the genetic stasis phenomenon [1]. But when are these tools disinhibited? Evolutionary leaps demonstrably very often occur after mass extinction events, which are the result of an external stressor, like quick and massive changes of environmental conditions. As a result, the survival of the species is endangered, which in turn could activate the transposable elements to change the genome, in order to adapt to the changed external conditions [1]. A genome that changes itself using genetic tools in response to external stressors strongly contradicts Darwin's evolutionary mechanisms of random mutations and natural selection, but McClintock's experiments in the 1940s and 1950s could show that the activity of transposable elements is massively upregulated under a stressor like exposure to radioactivity.

A theory is an abstract construct, that on the one hand should describe real events in the observable world, while on the other hand be able to make predictions that will either be confirmed or disproven by observations. If the observations of reality stand in contrast to predictions of a theory, like a leaping instead of a steady evolution or genome changing tools instead of random mutations as evolutionary driving force, then the theory has failed and needs to be revised and either be adapted or replaced by a new theory. Of course, the basic statement of the Theory of Evolution that all life on Earth evolved from a common ancestor, is still valid. Just the postulated mechanism for this evolution is, under incorporation of new scientific observations, untenable.

However, the question remains: How does life evolve? The mere existence of transposable elements or the observation of their function, by itself does not explain how they are activated and orchestrated. They are but tools and only the right use of them can create adaptive changes to the genome. This apparently seems to happen, since only parts of the genome are being changed during an evolutive leap, while other parts remain completely untouched. But what could control the transposable elements and how?

III. Biological Superorganisms

A Superorganism is a biological organism that consists of individual cells, that each by itself could form an independent life form. The classical examples of this are the bee hives and ant colonies: Each ant for example, could by itself be self-sufficient living creature, but because of a division of labor, like gathering food or reproduction, only the whole colony has a chance of long term survival and procreation by creating new queens to found new colonies. An essential condition for the functioning of a Superorganism is the ability of single cells to communicate with each other. Otherwise vital processes cannot be coordinated. Ants use pheromones to orchestrate all kinds of processes and as a result, they being individually neither very smart nor big, create something bigger, a hive, a Superorganism.

But in principle any multicellular lifeform can be defined as a Superorganism: A human being, for example, consists of billions of cells that could be independent life forms on their own, as bacteria. But through specialization on certain tasks, like bone, liver or neural cells, they are only able to survive in the collective of the Superorganism. To survive, the human body has evolved a lot of communication pathways to orchestrate vital process with its cells, e.g. hormones, neurotransmitters or electric signals. Together, the cells make sure that the organism survives and in summation, they create something much bigger: The conscious human being. This is quite remarkable, since the synchronized communication of billions and billions of nerve cells are able to store and process information in such a manner that a consciousness is created.

The notion of a Superorganism could even be extended:

Mankind itself can be seen as an organism, where each human takes on a specialized task, while communicating with each other and coordinating the functioning of mankind as a whole and therefore creating something much bigger.

In nature there are yet more inconspicuous, however not less fascinating, Superorganisms: Bacterial colonies [2] [4] [16].

The on the first look unimpressive bacteria were long seen as solitary life forms, that hardly do anything more than taking up food and multiplying. However, bacteria form highly complex colonies that are basically organized like multicellular lifeforms: Different cell types develop in order to perform specific tasks, like taking up food or disposing of waste products. Thereby they are creating colonies that can have 100 times more bacteria, than there are humans on Earth. To coordinate the individual cells efficient communication strategies are needed. There are, as in humans, different methods to transfer information and manage the behavior of the colony, e.g. Quorum Sensing, chemotactic signals, plasmid exchange, as well as communication through electric impulses via ion channels [16]. Quorum Sensing is a method for bacteria to measure the concentration of cells in the colony: Each bacterium produces signal molecules, that the other bacteria can sense using receptors on the outer cell membrane. This is for example important in the generation of a biofilm in order for the bacteria to protect themselves from an immune system. If too little bacteria are producing biofilm molecules, the immune system can wipe them out with a response, however if enough bacteria are present and they all start to produce it at the same time, the biofilm can protect the colony from the immune response to some extent. But chemical signals are also used in a direct information exchange between two bacteria: They can give up "pheromones" to "negotiate" with other bacteria and if successful, create plasmid-bridges to exchange genetic information, e.g. for antibiotics resistance, which in turn increases the number of antibiotic resistant strains [2]. During this negotiation, the recently discovered communication via electric signals probably has its role too [16].

All these communication methods suggest the conclusion, that bacterial colonies are capable of a similar level of self-organization and complexity as multicellular life. The coordination of a trillion single bacteria for a greater cause requires quite a lot of sophistication in the process. Since bacteria have been very successful life forms for billions of years, this task seems to have been resolved magnificently.

Depending on the availability of food, the size of the colony, the existence of enemies and many other factors, bacteria can coordinate their behavior, as in which genes to de/activate to for example synthesize toxins, or to hibernate or move as response to a threat.. But this is still only the tip of the iceberg in the bacterial adaptive repertoire: A series of experiments have shown, that bacterial colonies, which are subject to new, potentially deadly environmental conditions, e.g. antibiotics or a unusual culture medium, can change their own genome to adopt to the new conditions [1] [2] [4]. As early as 1984 Shapiro conducted an experiment with genetically changed bacteria [17]: Through deletion of one single letter in the genetic code, the bacteria could not use the nutrient medium as food anymore. The colony was on the verge of dying, but after two days the bacteria started growing again. Somehow the bacteria had managed to generate an adaptive mutation. Since such a short time span is not nearly enough to create such a specific mutation by chance, the logical conclusion is that the bacteria creatively adapted their genome to the new circumstances.

This adaptive evolution is in stark contrast to the predominant view of random mutations as the source of all evolutive changes and it raises the question of how bacteria could manage to provide the necessary computing power for an adaptive solution finding process.

One could also ask how an analogous process would work in the central nervous system of a multicellular organism. Be it the decision of a bird, where to build his nest, or of a baboon if it is worth it to fight for a higher social rank and risk injury, or an engineer developing a new automotive technology. In each of these cases, the preconditions of the problem have to be registered and analyzed, in order to develop creative action patterns. These then come under scrutiny to assess the potential capability and probability of success in solving the problem, after which a decision has to be made to choose the behavior that has the most likely chance of success. The brain uses for the processing of this (any?) kind of information neural activation patterns [18].

How it functions in detail is not understood completely, however it can be approximately imagined like this: If, for example, a primeval human ancestor happened to come across a Tiger in the wild, different sorts of data gathering, processing and action patters would emerge. First of all, the visual pixels the retina transmits to the brain need to be analyzed for patterns, in order to recognize and classify known objects. If the pattern the tiger's reflection of light leaves on the retina is compared to "predator-detection-patterns", e.g. the size, the form, the distance and a lot of these attributes match (the predator is very large, close and has big teeth), than a kind of emergency signal is send to ancient parts of the brain, where fight or flight circuits are activated. They in turn release signal molecules in form of neurotransmitter and hormones, like adrenaline, which change the behavior of the organism, by putting it in a state in which it is prepared to either stand its ground and fight, or take a viable option to escape. Therefore, blood sugar is raised and the blood flow of muscles increased, to increase their working capacity, while other bodily processes which are not needed for immediate survival are inhibited, e.g. digestion. In the brain, Noradrenaline and Adrenaline causes the neural cells to be easier excitable by electric or chemical stimuli, which lead to a more efficient signal transduction and information processing.

The body and neural system are now in a prepared state of high alertness, but there still needs to be a decision on how to react to the threat. Hence, the brain creates a big number of action patterns, which are in turn assessed for their probability of success. If the human is carrying a stone or a spear, it might be viable to fight. "Experience-patterns" from previous encounters can be incorporated, e.g. if the throw of a stone has scared hyenas away in the past, or a spear can only kill a big animal if it hits right and the probability of success can be evaluated by comparing similarities and differences of the previous and current situations. The action patterns for flight are being evaluated as well, but tigers can run faster and climb better than humans, so they do not have a high chance of playing out in favor for the human. The brain permanently produces new creative patterns which are then immediately evaluated and can be combined to new patterns if they are not good enough by themselves: The throw of a stone (aggressive behavior) could be combined with running away (passive behavior) when the tiger is startled. This pattern will be analyzed and if it seems sufficiently likely to be successful, it will be carried out and new patterns will be created to organize the actions. For this, the size and weight of the stone will be registered, as well as the distance and velocity of the predator, to create neural activation patterns for the motoric system in the brain. This then activates the right muscle fibers, at the right time, with the right strength, so the human throws the stone at the tiger as precisely as possible and starts to flee immediately after that. This all happens mostly unconsciously, without the human even recognizing the tremendous feats of data processing in the brain necessary to save him.

Although it is not clear as of yet, how the creation and processing of patterns in the brain is working in detail, it obviously has the capability to perform tremendous feats in fast data processing and creative solution development. And in addition, it can even create a consciousness.

In bacterial colonies, similar pattern based data processing methods could be used. Thereby the recognition, processing and creation of activation patterns could be done by communication between cells, or the through activation patterns inside the genome. Similar to the brain, where neurons can decipher information provided by sense organs to gain knowledge on the state of the surroundings, bacterial cells can do so through receptors on the outside of their cell membranes and communication loops with their neighbor cells [16], or through genome activation patterns (or a combination of both).

If a bacterial colony happens to be placed on a culture medium which it cannot break down, the surroundings are monitored through the receptors on the cell membrane in order to establish the chemical properties of the medium. If the usually consumed nutrients are missing, genes for digestive enzymes will be deactivated and the growth as well as the cell division processes will be halted to conserve energy. Depending on the soil the bacteria will pursue different strategies: If they are able to move on the soil, agile cells will be build, to swarm out and find nutrients. If they cannot move, this action pattern will be dismissed and the bacteria will stay where they are. Like in the aforementioned experiment by Shapiro, after some time passes, specific mutations will occur in the bacterial genome, which enable them to digest the unusual nutrient soil. Ben-Jacob et. al. concluded, that the bacteria needed two days' time to identify the underlying problem and devise a solution [2]. A similar experiment carried out by Hall, in which the bacteria needed two mutations to be able to survive, supports this theory: The bacterial colony needed twice as long to implement the needed mutations in their genome, which points to a twice as long "problem solving period" [19] [20].

But how can such an adaptive process work?

IV. Superorganisms as evolutionary driving force

Although it is known, that bacteria can adapt to a changing environment with specific adaptive genetic mutations, the in detailed process behind has not been resolved yet. How can changing environmental factors and subsequent information processing through pattern formation by single cells result in a change in the gene pool of the bacterial colony? Right now, this question cannot be answered definitively, just as the question of how a human creates actions from ideas. Still, important steps in this process are known, e.g. that the gathering and processing of information about the environment through pattern formation and processing between neural cells alters the body chemistry and leads to the execution of the desired action through the muscle cells. In bacterial colonies, the process is presumably pretty similar, although they by design have to use different instruments than muscles to adapt to or influence their environment. But just as human nerve cells can create action patterns which activate muscle cells to move, bacterial colony cells can generate patterns to activate transposable elements to perform specific actions in the genome. The process of adaptive genetic changes probably works something like this: The bacterial colony is exposed to a changing environment, in which it cannot survive with the current genetic repertoire. The environment is sensed through chemical receptors on the outer cell wall and as a first response, chemical messengers will be released to (de)activate existing genes, e.g. for growth and cell division, as a response to missing nutrients. These are epigenetic adaptions, which are reversible and used for short term adaption. Simultaneously the bacterial cells communicate with each other through chemical messengers and electric signals, similar to the neurons in the human brain. But how can communication between bacteria lead to the activation of transposable elements? Since uninhibited genome changing tools would wreak havoc on genome integrity by changing it all the time, mechanisms are needed to influence transposon activity in the cell: This is accomplished for example through changing enzyme activity, genome reading rates, as well as molecules, for example in multicellular organisms using microRNA or piRNA [21] [22]. These are short RNA-Chains which are only a couple dozen bases long. Due to their shortness, they were considered to be of little importance in the overall functioning of the cell processes, but it was discovered, that they are indeed vital and have many very important tasks, e.g. they are responsible for the genetic stability by inhibiting the activity of transposable elements [1].

In bacteria the mechanism which regulates the activity genome editing tools has not been discovered so far, but most certainly a similarly sophisticated inhibition process must exist.

These processes make sure that the genome content is usually stable and not subjected to changes, so the cell can rely on its genetic repertoire, which is adapted to the environment it lives in. However, if the environment changes dramatically and in a way which makes survival improbable or even impossible, this inhibition of transposable elements can be stopped and the genome can be changed. But instead of random changes all over the genome, changes happen only in the areas which are vital for the adaption to the new environment [1]. This means, the cell does not give up control of the transposons; it just regulates their activity in a way which makes them edit only certain parts of the genome. This may seem unbelievable to supporters of the darwinistic evolutionary model of random mutation, but decades of research proves that such a directed gene editing is in fact true. Such a control mechanism for creating localized mutations is essential for a Superorganism, since the parts of the genome, that are vital for the basic cell functions must not be changed, in order to survive. That this is possible has been shown by Shapiro's experiments: Otherwise there would have been changes distributed randomly all over the genome. However, only the areas necessary for survival were undergoing changes, while the rest was genetically stable.

Although the mechanisms have not been researched extensively, the communication of bacterial cells with each other probably leads to pattern creation similar to neurons [16]. These patterns are then evaluated and if a pattern is created, that can create genetic mutation pattern with a high likelihood of survival, this communication pattern could be translated in special molecules that orchestrate the activation of transposable elements in the desired way, as to change the genome in the desired way. Although the process is in reality much more complicated, in a simplified way, the directed genome editing process should work something like this. There are other factors that can lead to changes in the genome, for example viruses or the implementation of genetic material received by other bacteria through plasmid connections. But in the adaption to life threatening environmental changes, the Transposon-Pathway is likely the most important one. However, this does not mean, that the "gene-stealing" of parasites, for example the lamprey, which "stole" from his host a transposable element [23], or viruses that are able to take genes from blue whales and insert them into monkeys [1], are not important for genetic exchange between higher organisms. The horizontal gene transfer, meaning between cells/organisms/species instead of vertically to descendants, is much more common in bacteria than in higher multicellular organisms. One way to exchange genes is through the aforementioned plasmid connections, where a kind of tunnel is established between to bacterial cells, through which genetic fragments can be exchanged. Therefore it could be possible, that a colony "bombards" one single cell with genome changing signals and if the cell can afterwards survive and grow in the new environment, the changed genome parts can be distributed among the colony. Unfortunately, due to the darwinistic dogma that mutations only occur randomly, there has been no research in this field. One can just hope that there will be more research in the future. Like always, the reality will probably be much more complex and complicated and many new questions will arise.

But if bacterial colonies are ascribed a similar data processing capability as the human brain, and this really seems to be the case [16], many interesting conclusions can be drawn from this assumption: Bacteria seem to be able to grasp complex situations and perform even more complex changes in their genome. Many smart humans have been working on such a complex task, but so far only minor changes, e.g. the implementation of parasite resistance genes into agricultural plants, have been done successfully. Any more invasive genetic changes have so far not been successful, due to the complexity of the challenge. Bacterial colonies implement such changes seemingly with ease. This could of course mean that a tremendous data processing ability and effort is needed in order to perform useful genetic changes without damaging the functionality of the whole genome. Still, there is an even more impressive feat the communication between neural cells accomplishes in the human brain: the creation of a consciousness. Considering the incredible creative adaptability of bacterial colonies, suggesting some kind of "will to survive", it is not far-fetched to attest an, at least rudimentary, kind of consciousness to bacterial colonies [2]. This might be hard to acknowledge for many scientists, since just a few decades ago even higher multicellular organisms like animals were denied to have some kind of consciousness. Therefore it could be hard for many people to accept that something as abstract and unremarkable as a bacterial colony could be conscious. Since humans do not seem capable of even defining what the essence of a consciousness is, it is hard to apply it to "unconventional" life forms. In the natural sciences and philosophy, consciousness is often described as something that "perceives and experiences the environment" or "has thoughts". Still, all the current definitions cannot define the essence of a consciousness, which is astonishing since every human can experience his or her own consciousness most of the time (when awake). Maybe the intelligence of the humans is not enough, to really grasp the concept of a consciousness, which means one is limited to describing the properties it has. However, this makes it hard to prove objectively if a life form has a consciousness.

In the past, it was often alleged, that only animals which could recognize themselves in a mirror are truly conscious. This would imply that every consciousness would have an evolutionary advantage of recognizing itself in a mirror, which is obviously not the case. In fact, having a conscious and being able to project it into something of the "outer world" are two totally different abilities.

A better suited property to establish if a life form has a consciousness could be the existence of an intrinsic "will": Only a consciousness can want something. A computer can be programmed to go through calculation steps in order to keep itself alive, but only a life form can want to survive. This might be the single crucial benefit a consciousness has for a live form: Being motivated to survive and procreate could dramatically increase the chances of either happening. Also, since a consciousness is subjective, it is possible to program "good" and "bad", as in pleasant and unpleasant. A consciousness will strive to stay in a pleasant state (fed, warm etc.) and to avoid unpleasant states (pain, death). Such an advantage would explain the immense effort it needs to create a consciousness.

Bacterial colonies also have this condition for having a consciousness: They want to survive by any means necessary. The tremendous creative genetic-engineering feats the bacteria do on a daily basis cannot be explained any other way. They do not behave like a computer, which just behaves according to his predefined programming, but like the brain, which is able to deal with paradoxes. Ben-Jacob defined a paradox as being a new situation which has no known solution, in contrast to a problem, which has a known solution [2]. In the case of genetic adaption, a problem would be for example a high UV radiation, e.g. if a human from the northern hemisphere is exposed to the tropical sun. The problem of UV radiation is known and the skin cells "just" have to activate the genes for the production and implementation of melanin in the skin in order to become tanned and protected against harmful radiation. A paradox on the other hand experienced the bodies of the first humans who left Africa and settled in northern places: Since there was no regulation mechanism in place, their skin was permanently dark, which kept sunlight from reaching deeper skin levels, where it is needed in the right dosage to create Vitamin D. In the weak winter sun, a permanently dark skin could lead to a Vitamin D deficiency, which can be detrimental to overall health. For this paradox situation, there was no "master plan" or known solution, so it had to be solved creatively. A genetic adaption happened, making northern humans permanently white (a UV protection mechanism was added later, becoming brown (melanin) in Caucasian populations and yellow (keratin) in Asian populations). An unknown nutrient soil is also a paradox for a bacterial colony, for which it has to find a creative solution.

A view of nature, which attests bacterial colonies to have a consciousness, has of course far reaching consequences, just as an evolutionary theory, which is based on the existence of superorganisms, whose single cells communicate with each other and thereby process information and thereby create creative adaptive changes in the genome. Joining these two theories together, a species could also be seen as a superorganism. In this case, for example a human could be a single cell in the superorganism of mankind. The communication between single cells could happen through pheromones, genetic recombination through procreation or other chemical messenger molecules, which would lead to different genetic activation patterns and therefore other "output". Such a theory could also explain much of the evolutionary observations, e.g. how new species arise in the midst of their genetically static members of their species, as has happened quite often (despite Darwin's postulation, that only spatial isolation would lead to the evolution of new species) [1]. Part of a species could go into a "genome changing mode" due to changing environmental factors, resulting in changed genome activation patterns and creating microRNA which leads to specific changes in the genome, creating a new species. If this microRNA would only be produced for a short period of time, the spatial range would be restricted, which would lead to only a part of a population becoming a different species. However, it takes some time for such an evolution to occur (although much less than with Darwin's postulated mechanism) since the genome of multicellular organisms is much more complex than a bacterium's, which would lead to a longer time "to process data and devise a solution". Also, the distance between such a Superorganism's single cells would be pretty big and the communication rather slow. In contrast to the consciousness in the human brain, in which neural cells communicate with electric impulses that can be sent at a rate of up to 2000 per second, thereby enabling very fast perceiving, integrating and processing of information, a Superorganism whose "nerve cells" are single multicellular organisms, would have a much "slower" consciousness. This could of course be sped up, if each "single cell unit" would pre-process information beforehand, e.g. in form of gene activation patterns and resulting microRNA, which then just would have to be exchanged with other organisms. That this is possible and happening in nature has been proven by researches who studied the effects food has on gene activation patterns in humans. Researchers were baffled by the observation that rice consumption increases the cholesterol level in humans. A closer evaluation of this phenomenon revealed, that rice, although it hardly contains cholesterol, contains microRNA which blocked the gene for a cholesterol filtering receptor on the outer cell membrane. This was again baffling, since microRNA was expected to be digested in the stomach, long before it could enter the blood or even the cell core to alter gene activation. It was discovered however, that the microRNA was contained in an indigestible bubble, which could diffuse through the stomach walls into the blood stream and was even able to enter the cells and could thereby change gene activation rates in the DNA [24].

This mechanism could be one of the fundamental communication pathways for a Superorganism, since such microRNA bubbles could even be transmitted through diffusion by touch of the skin or maybe even through air. This still would be a slow communication pathway, which could explain why evolutionary leaps still took some few thousand to million years to occur after an environmental catastrophe.

However, the existence of such superorganisms would have much more far reaching consequences for mankind than a new and improved evolutionary understanding.

V. A Global Superorganism

If such biological Superorganisms exist on a small scale as bacterial colonies, they might as well exist on the large scale. On the one hand, one could see a species as a superorganism, which is in competition with other superorganisms (evolutionary arms race), always trying to outsmart their respective predators, prey, parasites or hosts....

On the other hand considering the full evolutionary history of live on Earth leads to an even more extreme hypothesis: Since all biological cells in the world today have a common ancestor, which is called LUCA (Last Universal Common Ancestor) and is presumed to have existed about 4.1 Billion years ago, it is very possible, that the whole biosphere is one giant Superorganism, spanning the whole Earth. One reason that speaks in favor of this hypothesis is the overwhelming similarity of today's biological organisms: Even though they have vastly differing ecological niches, metabolisms, habitats etc., the genetic code is still universal and the basic biological cell is still pretty much the same: Every cell has DNA and/or RNA, which could be influenced by microRNA or similar messenger molecules, therefore enabling communication between very different organisms (like humans and rice for example).

Bacterial cells which make up the biggest part of biological cells on Earth do not use microRNA; instead they have a much more efficient way to "communicate genetically": They use the aforementioned plasmid tunnels to transfer genetic information packages.

Considering all the unsolved mysteries in the evolutionary history under the aspect of the Superorganism theory, some findings suddenly make much more sense. The Cambrian Explosion, the starting point for all present multicellular organisms happened 550 Million years ago and is still one of the great mysteries of Evolution: Before this event, there were practically no multicellular organisms whatsoever for over 3.5 billion years, but in a matter of 5-10 Million years, all ancestor species of present multicellular life forms came into existence, hence the term "explosion". This is on the evolutionary scale a relatively small time span, definitely not enough to allow a random chance based approach of Darwin's natural selection to create such a huge amount of new and very complex species from single celled organisms. Evolution scientists still struggle very much to explain it. [1]

But instead of questioning the correctness of the existing evolutionary theories, which should be standard procedure in a scientific community if a model does not explain findings from the real world, it was declared a singularity. Such a procedure is highly unscientific. How had the modern understanding of physics evolved if the universal constancy of the speed of light would have been declared an unexplainable singularity and the accuracy of Newton's theory of mechanics would never have been questioned? Luckily, Einstein did question it and came up with the conclusion, that Newton's theory was inaccurate and would have to be improved, which led to the development of the theory of relativity, enabling a much more precise description of the real world. Something similar also needs to happen in the Theory of Evolution and the Cambrian Explosion. If geological findings from around the time of the Cambrian Explosion are considered, it becomes clear that it was a very unusual period of time with great stress for all life on Earth. For unknown reasons, the Earth became a snowball for three times in a relatively short time span, during which the surface was completely frozen and life had to survive under a thick ice layer below the ocean's surface [1].

"Relatively" short after the second global glaciation event, called the Marinoan Glaciation, the first multicellular organisms evolved. These were formless sponges without any symmetry whatsoever. An even greater evolutionary leap happened after the third glaciation: radially symmetric multicellular organisms evolved for the first time, which means a tremendous increase in complexity. A very sophisticated developmental process has to be in place to enable the growth of a multicellular organism with a defined body plan and different cell types from one single cell. The cells have to grow, divide and differentiate into the right cell type at the right time and in the right place. This incredibly hard task is coordinated with the Homeobox, short HOX, gene-complex [1]. This HOX-genes are activated chronologically one after the other, which leads to the de/activation of many genes, resulting in different cell types being created. The sponges were the first to have a "proto-HOX" gene complex, which lead to the cells staying attached to each other and creating a multicellular organism. During the evolutionary leap to radially symmetric organisms, this gene-complex was doubled and changed, creating more sophisticated HOX-genes, which in turn enabled the creation of very complex multicellular organisms. [25]

However, this was not the final evolutionary leap of the Cambrian Explosion: Only a few million years later another doubling, partially even quadrupling of the HOX-gene-complex happened, during which it was again changed tremendously, resulting in the creation of all ancestor species of present multicellular organisms with a longitudinal and a lateral body axis, or in laymen's terms: A front, a back, an up and down [1]. These methods of copying, inserting and changing the genome are tasks fulfilled by the transposable elements. But to create such an enormous evolutionary leap to incredibly complex multicellular life forms in such a "short" timespan, they would have to be controlled very precisely. Remarkably, the body plans of the present life forms has not changed much, all the nerve/circulatory systems, muscles, organs etc., all the important cell types making up "modern" species came into existence during the Cambrian Explosion. The German biologist Ernst Mayr formulated a comparison between modern and past multicellular life as "just a change in proportion, changing, melting or loss of some attributes which do not change the fundamental morphological plan" [1]. But how is it possible, that all multicellular predecessors of modern species developed in a few million years, after there was just single celled life for billions of years?

The hypothesis of a Global Superorganism could explain the Cambrian Explosion: Such a world spanning organism would be hit very hard by a global glaciation event, up until the point of total extinction. Maybe the Superorganism came to the conclusion that life on Earth is always threatened by unforeseeable events, e.g. meteorite impacts or gamma ray bursts. The logical consequence such a life form would deduct is that macroscopic multicellular organisms (or even ecosystems) would be much better at surviving such an event's consequences such as cold; maybe they could even change their environment, or as a last resort, bring life to another planet. Such a Superorganism would be able to devise the extremely sophisticated and complex body plans of such multicellular organisms. This task could hardly be any harder, billions of cells have to communicate, interact, process information, de/activate the right genes at the right time to form such an organism and afterwards enable his survival. But just creating multicellular life forms is not enough to save life on Earth, since the environment can change rapidly, to which the life forms would have to adapt quickly and creatively. Also, the first multicellular organisms were not able to bring life to another celestial body or solar system, an evolutionary process creating ever more improved organisms was needed. So the Global Superorganism devised a system of competition between "lower" Superorganisms (species). These had to compete for habitats, food, or were in a predator-prey relationship, prompting the famous evolutionary arms races. These species living in a kind of turbocapitalistic system with eat or be eaten rivalry leads to a fast innovation pace. After some time equilibriums get established in the form of functioning and robust ecosystems, but still small innovations happen in species continuing the arms race. If a major environmental stressor arises, such as a meteorite impact, after which often most of the species perish, the surviving Species-Superorganisms are forced to innovate and implement massive genome changes or die. As a result, often not a single new species arises, but instead a bunch, which can fill out the environmental niches freed up by the mass extinction [26]. In a way, this superorganism has found a way of "radiating" into sub-Superorganisms, which will then again compete and improve. This radiation is comparable to the human economic system: Nobody can predict which product will be successful at the market or be the best fit for the customer, especially since there are many different applications. Through the competition of rivaling corporations, different products get developed and produced, which then compete on the free market. Through the competition of Apple and Samsung Smartphones technical evolution happens, since each company wants to outsmart the other and produce a better product. As a result, the "Superorganism" of Mankind profits from technical advancement and better phones. There also is of course destructive competition, e.g. in wars, which create an extremely high innovation pressure, but also cause a lot of destruction. The Cold War led to the development of space flight technologies, but might as well have destroyed all of mankind through nuclear warfare. Such extreme competition rarely happens in nature, since single species would not evolve to such a great extent at once, as not to overhunt their prey and drive them to extinction. A prey species should also not evolve too much, as not to elicit a strong evolutionary response of its predator. Man, as often, is an exemption, since the great brain and intelligence advantage enabled the hunt of any big mammal, reptile or nautical species, often driving them into extinction in the past and present.

The trend to an accelerated innovative response can also be seen after a mass extinction event. The principles are the same as with the bacterial "evolution" in Shapiro's experiment: The environmental change caused the life form to not be genetically fit for long term survival anymore. After some latency, adaptive mutations arise, which enable survival again. The time it takes to create these mutations is proportional to the complexity of the needed mutations. For this reason, it can take some Millions of years for new species to arise after a mass extinction event, but it only takes days for bacteria to change a single gene letter [17].

According to supporters of Darwin's theory, evolution does not pursue a higher goal, it just is "pushed by chance" in a random direction, which might or might not be beneficial. According to the Superorganism Theory, Evolution does indeed serve a purpose, namely the creation of better life forms, which can sustain life on Earth (or elsewhere) after extreme catastrophes. This is a self-purpose for the Superorganism, since a consciousness "wants" to survive. The arising of humans could be considered the "goal" of Evolution, because of their unique intelligence and thirst for knowledge and control: This characteristic traits led to an evolution of human civilization that has arrived at the point where it became the first spacefaring species, which could potentially sustain life elsewhere in the universe, if Earth becomes uninhabitable. This would also explain our fascination with space itself, which is reflected in the amount of resources spent on the observation, development of space machinery, as well as the cultural approach of space themes, which all qualitatively and quantitatively far exceed for example the resources spent on the deep sea. The fascination of books, movies and documentaries of space fare, extraterrestrial lifeforms etc. reflect this trend as well, which might as well just exist, so humans bring life to other planets for the sake of the survival of the Global Superorganism.

One of the most important reasons for the existence of a Global Superorganism is the genetic component of "faith" Scientists found out that there is a genetic basis for having a belief in a higher power/being [27] . This has baffled scientists, since such an "instinct for belief" has no direct advantages. It could of course have an advantage for the creation of a group mentality in a tribe or community, but in the modern civilization all the classical functions of religion have been taken over by other organs, for example by jurisdictional or educational institutions. Group mentality is also created by nations, corporations or other groups, philosophy and ethics are perpetuated through societal norms without the need of a central religious entity. The only feature that is uniquely found in religion is the belief in a "higher power" or "being" as well as giving life a "sense". However, this belief instinct has caused billions of people to think about higher powers, beings and their own existence. One possible explanation for this could be that the Superorganism has left a hint at its own existence. This could mean that the "God" humans have searched for since the beginning of their existence, could be the biosphere.

VI. A view on the World

The existence of biological Superorganisms, especially a Superorganism consisting of the entirety of Earth's biosphere, would change the view of humankind on nature, its role in it and life itself dramatically. Every living organism on Earth would be part of something greater. Every human, every animal, every plant, even every bacterium would therefore be part of a giant Superorganism and do its part for the "greater good": The survival of life itself.

The Superorganism theory enables a new grasp on a much more complex and fascinating world in which every bacterial colony, every species and even nature itself are Superorganisms. The whole Earth is spanned by a vast highly complex network of consciousnesses that interact with each other, exchange information and compete or cooperate. This may sound unimaginable, but considering the extremely long time span of over 4 billion years during which life had time to develop to the highly complex biosphere it is today, the existence of this complex living nature is still a miracle, but not inexplicable.

Still, the question arises, of how it all came to be in the beginning. How could such a Superorganism develop, that made the evolution of the complex biosphere possible? What happened in the downright magical timespan when inanimate matter on Earth became a conscious living being? According to research of the last few decades, life developed on underwater Vulcans, also called black smokers. These volcanic eruptions set free a stream of molecules and energy, which in turn led to high concentrations of complex molecules like RNA developing around these smokers. Since interaction between chemical molecules can create a consciousness in the human brain, such a feat can also be possible with the highly complex molecules that were created at the black smokers. Still, the emergence of life is even by modern scientific standards a "miracle", since the reductionist scientific approach cannot explain how matter in an inanimate universe can become alive and even conscious. This has long been the basis for an argument between scientists and philosophers, as the existence of life and consciousness can, according to philosophers, only mean that there is an inherent physical property in the universe that enables the existence of such phenomena. Still, even if RNA interacting with each other might have formed a consciousness, this still does not explain how complex cells came to be, since a consciousness without the means to influence its environment is doomed for extinction. This means that parallel to the emergence of a consciousness, some kind of rudimentary genetic "tools" must also have come into existence, which the consciousness could somehow "operate". Both phenomena should each have a very low probability of happening, but the fact that life emerged pretty soon after Earth had cooled of enough to have a (mostly) solid crust, tells another story. It seems like the probability for life to emerge, if the circumstances allow it, cannot be so low. It can be argued that this would be a statistical process and it happening so fast means it cannot be too improbable. If this is in fact the case, life on Earth could have emerged a few times. Even on other planets with the right preconditions, like temperature, available molecules and elements, as well as some kind of energy source, life should emerge rather quickly. And there are countless celestial bodies in our galaxy which should have such conditions; some of them are even in our solar system, e.g. Mars or some moons of Jupiter. This could mean that a close examination of these places could detect extraterrestrial life in the near future.

Although the existence of a global Superorganism is by itself incredibly interesting, it has much more far reaching consequences for mankind, that perpetrate every aspect of human life, be it moral, social or economic For the functioning of a global network of Superorganisms a finely tuned ratio of cooperation to competition needs to exist: Although the cooperation of single actors with each other increases efficiency and enables better problem solving capabilities, competition is vital to motivate innovation. This is analogous to human behavior: Mankind has optimized its ratio of competition to cooperation in the last millennia. Early on a strong competition between tribes led to humans searching for new habitats and developing new survival strategies. The former led to mankind expanding all over the globe, while the latter led to innovations like agriculture and new hunting/war weaponry. Over time civilizations were established by the most successful tribes, which enabled more and more humans to cooperate with each other, resulting in ever greater numbers of individuals forming communities with an ever lesser degree of relatedness. Therefore new methods were needed to create a sense of community, so humans could feel like part of something bigger and be motivated to pursue goals for the "common good". As a result, institutions like religions or states were introduced that replaced tribal affiliation. However, these higher level institutions again had to compete with each other, resulting in warfare to expand kingdoms, spread religion or improve economic positions. Wars were always times of extreme levels of competition, resulting in a high pressure to innovate and cooperate within the respective institutions. This yielded ever more sophisticated strategies and weapon systems, making wars ever more destructive, reaching the peak in the 20th century with the World Wars and the Cold War. Paradoxically, the creation of more destructive weaponry led to wars being less profitable in any way, up until the point where a war would mean Mutually Assured Destruction [28]. As a result competition was shifted from military warfare to economic rivalry: The US and Japan don't fight each other with soldiers and bombs anymore, but instead with innovative products, legislation and currencies. Systems for economic competition however, need to be devised and carefully controlled in order to guarantee the desired ratio of competition to cooperation. Of course today's economic and political systems, which are ingenious inventions, still have many shortcomings, but considering the relatively short time they exist and their enormous complexity this is not surprising. All the current problems, e.g. extreme concentration of wealth and poverty, corruption of politicians by influential industries etc. can be attributed to "childhood diseases" of the system, which probably will be cured in the next centuries.

Interestingly, the strategies mankind uses with capitalism are pretty close to what is used in wild nature. Different actors (species/humans) either cooperate (symbiosis/community building) or compete (predator-prey/corporations), while serving a "higher purpose": The competition between species results in superiorly adapted life forms that can sustain life on Earth better (and thereby the Superorganism). For humans the competition between corporations and states results in technical advancement which ensures the survival of mankind (and thereby also the Superorganism). Just as in nature, this is more a pseudo-competition: The corporations fight to survive (and the humans for an income) by creating and selling innovative products, but the real profiteer is always mankind as a whole. New technologies enable it, to beneficially influence its probability for survival, e.g. by predicting and (hopefully) weakening climate change or by preventing a giant asteroid from destroying the whole world in the future. In doing his work, each human is serving a purpose in the functioning of the whole of mankind, its technical advancement and the survival of the Global Superorganism.

Humans, as (at least for now) only species with higher intelligence, have the responsibility to be aware of their power and use it wisely and morally, especially without destroying nature. Nature should always be seen as something "sacred", of which we all are a part of. Everybody is responsible, since every human is part of a community, many of which in turn form societies that interact with other societies. Mankind may be more than the sum of all humans, but each and every one can influence it. Of course some people have a higher influence, e.g. leaders of giant corporations or states, that can influence the development of whole societies, in positive but also in negative ways. These humans have an especially high responsibility, although this was often ignored in the past, when the pursuit of money and power was more of an end in itself. Therefore vast amounts money and incredible power were accumulated without bringing the "owner" or his descendants any more considerable benefits, but instead leading to overexploitation of humans and destruction of nature. Countless species have been rendered extinct, tracts of lands and rivers polluted and innumerable human lives destroyed, even the survival of mankind was (or even is) at risk, because humans succumbed to greed. Nevertheless, the past decades have brought (at least partly) a change of thinking: Instead of accumulating wealth just for the purpose of getting richer, more and more people ask themselves, how they can fulfill their social responsibility. Bill Gates has shown with his foundation that accumulated wealth can be used for the common good. Elon Musk has used the fortune he made from his innovative internet companies to create Tesla and SpaceX (among others) in order to solve humanity's problems and make it a multiplanetary civilization.

But money needs to be used sensibly to prevent it from being destructive. Often approaches to help people end up hurting them more, as it is the case with developmental aid. Donations of American and European food and clothes to Africa have wrecked local farmers, merchants and producers. Monetary donations for e.g. schools help dictators spend less on education and more on the military complex, which in turn stabilizes their position and makes it easier for corporations to overexploit nature and people through corrupt officials.

It would be more sensible to spent money to create production capacities, which in turn create long term jobs and wealth, instead of just dumping excess food, clothes and undirected money into these countries.

Although humanity has quite a lot of problems right now, which mostly stem either from too much concentrated capital, short term thinking in politics and economics and the fundamental attitudes of many humans, one can be hopeful that those problems will be solved with time. There are some statistical megatrends in human civilization that take place over very long periods of time: The propensity of violence has decreased for the last millennia, while intelligence and wealth has increased (although not too balanced). Mankind has left is "childhood phase" behind and has entered the "teenager-growth phase": As in a human such a growth phase creates a lot of conflict and growing pains with the environment but also one's self, since lacking maturity and experience leads to rash and subpar decisions. From the consequences of these decisions however, needs to be learned, so mankind can improve its ability to solve its problems. For this it is essential that any institution, system, law, theory or procedure is open to be analyzed scientifically. This can reveal and fix weak points, but only if nothing is exempt from analysis and reformation. In the near to medium distant future, mankind has good chances of survival and prosperity.

In the long term however, life on Earth is doomed. Life has been existing for over 4 Billion years, but in less than 2 Billion years the sun will destroy all life on Earth. For complex life the expected timespan of survival is much shorter, since in about 200-300 Million years plants will not be able to photosynthesize oxygen anymore [29]. This means life on Earth is already well past its half time and although the remaining time seems to be quite long, especially since humans have achieved a tremendous technological evolution in the past 150 years, the exploration of space and development space flight technologies should be of the highest priority. Giant asteroids could smash into the Earth and destroy all complex life. Since it took 500 Million years for humans to evolve from the first multicellular organisms, the remaining 200 Million years might not be enough to create another highly intelligent species capable of spaceflight, in effect dooming Life to die on Earth. This should also be considered with nuclear/chemical/biological warfare, it might destroy not only mankind but all of life in the long term. Establishing a "backup" in form of a Martian Colony would be very sensible. Also for the long term goal of expanding life to interstellar habitats, gaining knowledge and experience in space flight and the establishment of colonies, maybe even Terraforming, are vital. But until giant space ships will bring life to other stars, mankind has a long and hard, but also incredibly exciting way before itself.

Even more pressing matters in the present time are overpopulation, overexploitation of resources, and the destruction of whole ecosystems. To solve them, cooperation between populations needs to increase, as well as long term thinking. Another problematic trend in industrial nation is the disillusioning of people: It may seem to many people that a deeper sense of purpose in life has gone missing. This is probably also partly due to people having more spare time to think about their place in the universe, instead of needing to survive on a day to day basis. Also the easy "truths" offered by religions and their ignorant and arrogant approach to scientific discoveries have disappointed many. As a substitute many develop a tendency for a hedonistic lifestyle, filling their inner void with a series of moments with joyful experiences. This however cannot bring long term happiness, since the brain gets used quickly to such quick joys like alcohol or sweet sugar, so one needs higher and higher doses to feel "happy". This can result in excesses, moderation is passed aside; obesity, diabetes, addiction and depression can be the outcome. A similar trend could be observed in the late Roman Empire, where the wealthy social classes tended to have hedonistic excesses. Similar tendencies are developing today, where humans pursue more money and power, hoping the next increase in wealth or influence will satisfy and make them happy. However, a wealthy society does not have to be disillusioned; maybe just a different outlook on life is needed.

Many people have begun to prefer a fulfilling job and a well-balanced social and family life over the pursuit of material goods. If the question for the meaning of life is considered under the Superorganism theory, this change makes sense: If we have been "developed" by a Superorganism that lives through us, all action-patterns that would ensure this survival should make us happy, as to reinforce the habit. A fulfilling work that is seen as doing some useful, as in making sure mankind is able to function or advance further, as well as a balanced social and family life to ensure communication and procreation, can "incidentally" lead to a fulfilled content life. Especially because of the media's fixation on money and fame as the highest of all goals, it is often forgotten that (nearly) every job that is paid, serves a purpose: Corporations should only spend money on something that is needed, as in making a partial system of humankind function. Be it garbage disposal so the metropoles do not suffocate in their own waste, or driving a truck to make sure goods can be exchanged in the economy. Consuming too much media which suggests only "special" jobs such as actor or professional athlete, deserve very high recognition can cause frustration. Instead, everybody should be aware that even the most basic work is needed to make sure mankind is functioning and everybody is serving a purpose in the greater good.

It is often asserted, that the human mind is not capable of comprehending the meaning of life. More probable is however, that the meaning of life is far more trivial than is often assumed: The meaning of Life (on Earth) is that a Superorganism with a consciousness wants to live. The meaning of a (human) life therefore is making sure the survival of nature is guaranteed, through meaningful work, procreation and social exchange.

Another cultural aspect that has to be completely revised is the definition of "God". A God used to be a "supernatural" "all-knowing" and "almighty" higher being that not only created us, but also still controls everything. Why humans are searching for such a God has been a mystery for scientists for quite a long time. Studies suggest there is a "genetic instinct for faith", but the benefits of such an instinct were unclear. Other communities based on language, tribe or nation can just as well create a sense of group affiliation. So if humans already have an instinct to cooperate in groups, why is there an instinct needed to believe in a higher power/being? One explanation would of course be that the Superorganism has left a clue about its own existence, as to be "found" by humans. This could mean that "the God" people have been searching for and trying to explain since the beginning of mankind could be a biological Superorganism. Such a hint does make sense to make humans respect nature and prevent short sighted overexploitation and destruction.

This would mean that the definition of "God" would not have to be something esoteric and supernatural, but instead could be a "higher" living being, like a biological Superorganism. Such a "God" could be scientifically measured and described. This new understanding would also exclude any "supernatural influence", which means that everything that happens to us is not part of a bigger plan, but rather a mixture of chance and the consequences of our actions. This would make mankind mature, responsible and self-determined.

An atheistic viewpoint also makes humans responsible, but is for many people concocted with a loss of "purpose" or "meaning". But both extreme positions can have a negative impact on humanity: An esoteric dream of a perfect afterlife can make one see the present life as transient and of lesser importance. Overexploiting our planet or causing a lot of destruction would then not be such a big problem, since one still has the perfect afterlife. An atheistic viewpoint can lead to the nihilistic conclusion, that nothing has a deeper purpose, so a hedonistic lifestyle would make the most sense. Of course, these are extreme examples and most people of either side would fall somewhere in between.

But the way of thinking that our world is the only one we have (for now) and that we serve a deeper purpose in life, should be very beneficial for a sustainable advancement of humanity.

It could bring humans closer together, since every human is part of something bigger: The Global Superorganism.

References

[1] J. Bauer, Das kooperative Gen, Hamburg: Hoffman und Campe Verlag, 2008.

[2] E. Ben-Jacob, „Bacterial wisdom, Gödel's Theorem and creative genomic webs," *Physica A,* pp. 57-76, 31 August 1997.

[3] E. Bell, P. Boehnke, T. Harrison und W. Mao, „Potentially biogenic carbon preserved in a 4.1 billion-year-old zircon," *Proceedings of the National Academy of Sciences of the United States of America,* 2015.

[4] H. Bloom, Global Brain: The Evolution of Mass Mind from the Big Bang to the 21st Century, Wiley, 2001.

[5] J. Shapiro, „ 21st century view of evolution: genome system architecture repetitive DNA, and natural genetic engineering," *Gene 345,* pp. 91-100, 2005.

[6] J. Shapiro, „Genome Informatics: The Role of DNA in Cellular Computations," *Biological Theory,* pp. 288-301, September 2006.

[7] D.-Y. Jiang, R. Motani und J.-D. Huang, „A large aberrant stem ichthyosauriform indicating early rise

and demise of ichthyosauromorphs in the wake of the end-Permian extinction," *Nature,* 23 May 2016.

[8] K. Peterson und N. Butterfield, „Origin of the Eumetazoa: Testing ecological predictions of molecular clocks against the Proterozoic fossil record," *Proceedings of the National Academy of Sciences of the United States of America,* Mai 2005.

[9] S. Gould und N. Eldredge, „Punctuated Equilibrium Comes of Age," *Nature 366,* pp. 223-227, 1993.

[10] P. Morris, L. Ivany, K. Schopf und C. Brett, „The challenge of paleoecological stasis: reassessing sources of evolutionary stability," *Proceedings of the National Academy of Sciences of the United States of America,* 1995.

[11] B. McClintock, „The origin and behavior of mutable loci in maize," *Proceedings of the National Academy of Sciences of the United Sates of America,* pp. 344-355, 8 April 1950.

[12] S. Ravindran, „Barbara McClintock and the discovery of jumping genes," *Proceedings of the National Academy of Sciences of the United States of America,* 2012.

[13] I. H. G. Consortium, „Initial sequencing and analysis of the human genome," *Nature 409,* pp. 860-921,

2001.

[14] C. Darwin, The Origin of Species, 1859.

[15] R. Wesson, Beyond Natural Selection, 1993: The MIT Press, 1993.

[16] A. Prindle, J. Liu, M. Assaly, S. Ly, J. Garcia-Ojalvo und G. Süel, „Ion channels enable electrical communication in bacterial communities," *Nature,* 2015.

[17] J. Shapiro, „Observations on the formation of clones containing araB-lacZ cistron fusions," *Molecular and General Genetics MGG,* Bd. 194, Nr. 1, 1984.

[18] H. Beck, Biologie des Geistesblitzes - Speed up your mind!, Springer Spektrum, 2013.

[19] B. Hall, „Adaptive evolution that requires multiple spontaneous mutations. I. Mutations involving an insertion sequence," *Genetics,* 1988.

[20] B. Hall, „Adaptive evolution that requires multiple spontaneous mutations: mutations involving base substitutions," *Proc Natl Acad Sci U S A,* p. 5882–5886, 1 July 1991.

[21] S. Bhattacharyya, R. Habermacher, U. Martine, E. Closs und W. Filipowicz, „Relief of microRNA-Mediated Translational Repression in Human Cells

Subjected to Stress," *Cell,* 13 Juni 2006.

[22] W. Theurkauf, „Transposon Silencing of Small RNAs," *Developmental Cell,* 2011.

[23] S. Kuraku, H. Qiu und A. Meyer, „Horizontal Transfers of Tc1 Elements between Teleost Fishes and Their Vertebrate Parasites, Lampreys," *Genome Biology and Evolution,* 2 August 2012.

[24] „Spektrum der Wissenschaft," Spektrum Springer Verlag, 19 September 2011. [Online]. Available: http://www.spektrum.de/news/pflanzliche-rna-aus-der-nahrung-reguliert-koerperprozesse/1123634. [Zugriff am 27 Oktober 2015].

[25] G. Wagner, C. Amemiya und F. Ruddle, „Hox cluster duplications and the opportunity for evolutionary novelties," *Proceedings of the National Academy of Sciences of the United States of America,* 2003.

[26] D.-Y. Jiang, R. Motani, J.-D. Huang, A. Tintori, Y.-C. Hu, O. Rieppel, N. C. Fraser, C. Ji, N. P. Kelley, W.-L. Fu und R. Zhang, „A large aberrant stem ichthyosauriform indicating early rise and demise of ichthyosauromorphs in the wake of the end-Permian extinction," *Nature - Scientific Reports 6,* 2016.

[27] M. Blume, „Vererbte Religion," *Spektrum der Wissenschaft,* 2011.

[28] Strategic Studies Institute, U.S. Army War College, Getting Mad: Nuclear Muttual Assured Destruction, Its Origins and Practice, 2004.

[29] M. Hahn und D. W. Savin, „How to Survive Doomsday," *Nautilus,* 05 05 2016.

Notes

www.ingramcontent.com/pod-product-compliance
Lightning Source LLC
Chambersburg PA
CBHW070334190526
45169CB00005B/1882